# POWER ELECTRONICS FOR MODERN WIND TURBINES

Frede Blaabjerg and Zhe Chen
Institute of Energy Technology
Aalborg University, Denmark

*SYNTHESIS LECTURES ON POWER ELECTRONICS #1*

MORGAN & CLAYPOOL PUBLISHERS

## ABSTRACT

Wind energy is now the world's fastest growing energy source. In the past 10 years, the global wind energy capacity has increased rapidly. The installed global wind power capacity has grown to 47.317 GW from about 3.5 GW in 1994. The global wind power industry installed 7976 MW in 2004, an increase in total installed generating capacity of 20%. The phenomenal growth in the wind energy industry can be attributed to the concerns to the environmental issues, and research and development of innovative cost-reducing technologies.

Denmark is a leading producer of wind turbines in the world, with an almost 40% share of the total worldwide production. The wind energy industry is a giant contributor to the Danish economy. In Denmark, the 3117 MW (in 2004) wind power is supplied by approximately 5500 wind turbines. Individuals and cooperatives own around 80% of the capacity. Denmark will increase the percentage of energy produced from wind to 25% by 2008, and aims for a 50% wind share of energy production by 2025.

Wind technology has improved significantly over the past two decades, and almost all of the aspects related to the wind energy technology are still under active research and development. However, this monograph will introduce some basics of the electrical and power electronic aspects involved with modern wind generation systems, including modern power electronics and converters, electric generation and conversion systems for both fixed speed and variable speed systems, control techniques for wind turbines, configurations of wind farms, and the issues of integrating wind turbines into power systems.

## KEYWORDS

Control of wind energy conversion system, Grid integration, Power electronics and converters, Power quality, Wind farms, Wind turbines

# POWER ELECTRONICS FOR MODERN WIND TURBINES

Copyright © 2006 by Morgan & Claypool

Power Electronics for Modern Wind Turbines

Frede Blaabjerg and Zhe Chen

www.morganclaypool.com

1598290320    paper    Blaabjerg/Chen
1598290339    ebook    Blaabjerg/Chen

DOI 10.2200/S00014ED1V01Y200602PEL001

A Publication in the Morgan & Claypool Publishers' series
*SYNTHESIS LECTURES ON POWER ELECTRONICS*
Lecture #1

First Edition
10 9 8 7 6 5 4 3 2 1

Printed in the United States of America

# Contents

# Acknowledgement

The authors wish to thank their colleagues who have worked on Wind Turbine Research Programs in the Institute of Energy Technology, Aalborg University, including staff and students, for some of the results presented in the publication.

# Introduction

Wind turbine technology is one of the fastest developing renewable technologies. The recent development started in the 1980s with a few tens of kilowatt power rating wind turbines to today's megawatt range wind turbines. In the earlier time wind power production did not have any serious impacts on the power system operation and control, but now it plays an active part in the grid since the wind power penetration level is increasing rapidly. The technology used in wind turbines was in the beginning based on squirrel-cage induction generators directly connected to the grid. By that, power pulsations in the wind are almost directly transferred to the grid. Furthermore, there is no active control of the active and reactive power that typically are the control parameters to the system frequency and voltage. As the power range of the turbines increases these control parameters become more important. Also the introduction of power electronics has changed the basic characteristic of wind turbines from being an energy source to be an active power source [1]. With the price of the power electronic devices falling, the solutions with power electronics become more and more attractive.

This monograph will first introduce the basic electrical components and systems in wind power conversion systems, and then the generators and the development in power electronics will be briefed. Then various wind turbine configurations will be presented. Also some control methods will be explained. The grid integration of wind turbines becomes more important, and therefore will be discussed regarding the different characteristics of the various wind turbine systems.

C H A P T E R   1

# Wind Energy Conversion

The development in wind turbine systems has been steady for the last 25 years and four to five generations of wind turbines exist. The main components of a wind turbine system, including the turbine rotor, gearbox, generator, transformer, and possible power electronics, are illustrated in Fig. 1.1.

The turbine rotor converts the fluctuating wind energy into mechanical energy, which is converted into electrical power through the generator, and then transferred into the grid through a transformer and transmission lines.

Wind turbines capture the power from the wind by means of aerodynamically designed blades and convert it to rotating mechanical power. The number of blades is normally three and the rotational speed decreases as the radius of the blade increases. For meagwatt range wind turbines the rotational speed will be 10–15 rpm. The weight-efficient way to convert the low-speed, high-torque power to electrical power is to use a gearbox and a generator with standard speed. The gearbox adapts the low speed of the turbine rotor to the high speed of the generator. The gearbox may be not necessary for multipole generator systems.

The generator converts the mechanical power into electrical energy, which is fed into a grid through possibly a power electronic converter, and a transformer with circuit breakers and electricity meters. The connection of wind turbines to the grid is possible at low voltage, medium voltage, high voltage, and even at the extra high voltage system since the transmittable power of an electricity system usually increases with increasing the voltage level. While most of the turbines are nowadays connected to the medium

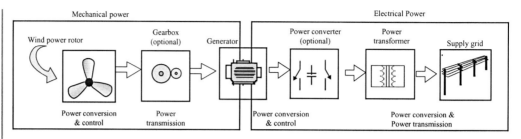

**FIGURE 1.1:** Main components of a wind turbine system.

voltage system, large offshore wind farms are connected to the high and extra high voltage level.

The electrical losses include the losses due to the generation of power, and the losses occur independently of the power production of wind turbines and also the energy used for lights and heating. The losses due to the power generation of the wind turbines are mainly losses in the cables and the transformer. The low-voltage cable should be short so as to avoid high losses. For modern wind turbine system, each turbine has its own transformer to raise voltage from the voltage level of the wind turbines (400 or 690 V) to the medium voltage. The transformer is normally located close to the wind turbines to avoid long low-voltage cables. Only small wind turbines are connected directly to the low-voltage line without a transformer or some of small wind turbines are connected to one transformer in a wind farm with small wind turbines. Because of the high losses in low-voltage lines, large wind farms may have a separate substation to increase the voltage from a medium voltage system to a high voltage system. The medium voltage system could be connected as a radial feeder or as a ring feeder.

At the point of common coupling (PCC) between the single wind turbines or the wind farm and the grid, there is a circuit breaker for the disconnection of the whole wind farm or of the wind turbines. Also the electricity meters are installed usually with their own voltage and current transformers.

The electrical protective system of a wind turbine system needs to protect the wind turbine and as well as secure the safe operation of the network under all circumstances.

For the wind turbine protection, the short circuits, overvoltage, and overproduction will be limited to avoid the possibly dangerous damage to the wind turbine system. Also the system should follow the grid requirements to decide whether the wind turbine should be kept in connection or disconnected from the system. Depending on the wind turbine operation requirement, a special relay may be needed to detect if the wind turbine operates in a grid connection mode or as an autonomous unit in an isolated part of the network due to the operation of protection devices.

The conversion of wind power to mechanical power is done aerodynamically as aforementioned. It is important to control and limit the converted mechanical power at higher wind speed, as the power in the wind is a cube of the wind speed. The power limitation may be done by stall control (the blade position is fixed but stall of the wind appears along the blade at higher wind speed), active stall control (the blade angle is adjusted in order to create stall along the blades), or pitch control (the blades are turned out of the wind at higher wind speed).

Fig. 1.2 shows the power curves of different types of turbine rotor power limitation methods [2]. It can be seen that the power may be smoothly limited by rotating the blades either by pitch or by active stall control while the power limited by the stall control shows a small overshoot, and this overshoot depends on the aerodynamic design.

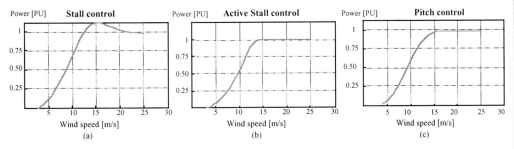

FIGURE 1.2: Power characteristics of fixed speed wind turbines [2]. (a) Stall control, (b) active stall control, and (c) pitch control.

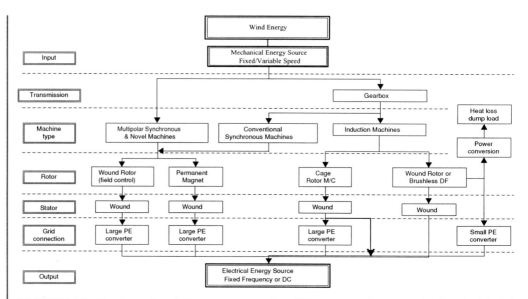

**FIGURE 1.3:** Roadmap for wind energy conversion. PE = power electronics; DF = doubly fed [3, 4].

The possible technical solutions of the electrical system are many and Fig. 1.3 shows a technological roadmap starting with wind energy/power and converting the mechanical power into electrical power. It involves solutions with and without gearbox as well as solutions with or without power electronic conversion. In the following chapters, main wind turbine configurations will be presented and explained.

C H A P T E R   2

# Modern Power Electronics and Converter Systems

Many types of wind turbines, such as variable speed wind turbine systems, use power electronic systems as interfaces. Since the wind turbine operates at variable rotational speed, the electric frequency of the generator varies and must therefore be decoupled from the frequency of the grid. This can be achieved by using a power electronic converter system. Even in a fixed speed system where the wind turbines may be directly connected to the grid, thyristors are used as soft-starters. This chapter discusses the modern power electronics, which play an important role.

## 2.1 POWER ELECTRONIC DEVICES

Power electronics has changed rapidly during the last 30 years and the number of applications has been increasing, mainly due to the developments of semiconductor devices and microprocessor technology. For both cases higher performance is steadily given for the same area of silicon, and at the same time the price of the devices is continuously falling. Three important issues are of concern in using a power electronic system. These are *reliability*, *efficiency*, and *cost*. At the moment the cost of power semiconductor devices is decreasing 2–5% every year for the same output performance. Fig. 2.1 shows some key self-commutated devices and the area where the development is still on going.

The only power device that is no longer under development (see Fig. 2.1) is the silicon-based power bipolar transistor because MOS-gated devices are preferable in the

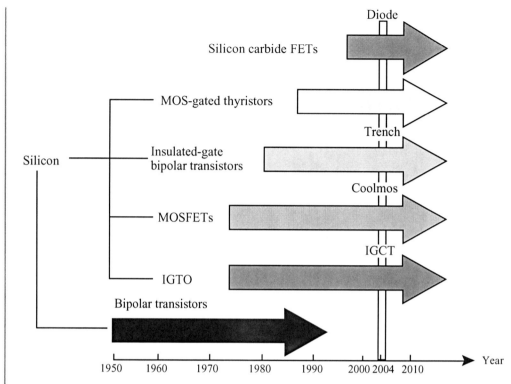

FIGURE 2.1: Development of power semiconductor devices in the past and in the future [5].

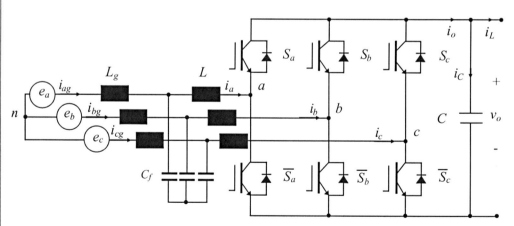

FIGURE 2.2: Circuit diagram of a voltage source converter (VSC) with IGBTs.

sense of easy control. The breakdown voltage and/or current carrying capability of the components are also continuously increasing. Also, important research is going on to change the material from silicon to silicon carbide. This may dramatically increase the power density of power converters, but silicon carbide based transistors on a commercial basis, with a competitive price, will still take some years to appear on the market.

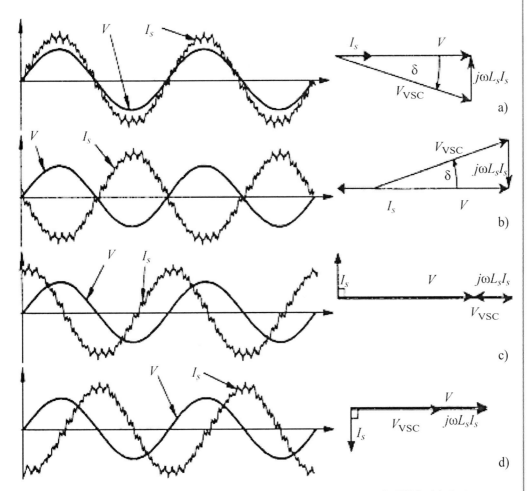

FIGURE 2.3: Waveforms of bidirectional active and reactive power of a VSC. (a) Active power flow from the ac system to the converter dc side. (b) Active power flow from the converter dc side to the ac system. (c) The converter generating reactive power. (d) The converter consuming reactive power.

## 2.2   POWER ELECTRONIC CONVERTERS

Power electronic converters are constructed by power electronic devices, driving, protection and control circuits. A converter, depending on the topology and application, may allow both directions of power flow and can interface between the load/generator and the grid. There are two different types of converter systems: grid commutated and self commutated converter systems. The grid commutated converters are mainly thyristor converters, 6 or 12 or even more pulse. This type of converter produces integer harmonics which in general requires harmonic filters [6, 7]. Also thyristor converters are not able to control the reactive power and consume inductive reactive power.

Self commutated converter systems are mainly pulse width modulated (PWM) converters, where IGBTs (Insulated Gate Bipolar Transistor) are mainly used. This type of converter can control both active power and reactive power [8, 9]. That means the reactive power demand can be delivered by a PWM-converter. The high frequency switching of a PWM-converter may produce harmonics and interharmonics. In general these harmonics are in the range of some kHz. Due to the high frequencies, the harmonics are relatively easier to be removed by small size filters. Fig. 2.2 shows a typical power electronic converter consisting of self commutated semiconductors such as IGBTs and Fig. 2.3 shows the waveforms of different operation modes.

CHAPTER 3

# Generator Systems for Wind Turbines

Both induction and synchronous generators can be used for wind turbine systems. Induction generators can be used in a fixed-speed system or a variable-speed system, while synchronous generators are normally used in power electronic interfaced variable-speed systems. Mainly, three types of induction generators are used in wind power conversion systems: cage rotor, wound rotor with slip control by changing rotor resistance, and doubly fed induction generators. The cage rotor induction machine can be directly connected into an ac system and operates at a fixed speed or uses a full-rated power electronic system to operate at variable speed. The wound rotor generator with rotor-resistance-slip control is normally directly connected to an ac system, but the slip control provides the ability of changing the operation speed in a certain range. The doubly fed induction generators provide a wide range of speed variation depending on the size of power electronic converter systems. In this chapter we first discuss the systems without power electronics except the thyristor soft starter, and then discuss the variable-speed wind turbine systems, including those with partially rated power electronics and the full-scale power electronic interfaced wind turbine systems.

## 3.1    FIXED-SPEED WIND TURBINES

In fixed-speed wind turbines, the generator is directly connected to the mains supply grid. The frequency of the grid determines the rotational speed of the generator and thus of the rotor. The generator speed depends on the number of pole pairs and the frequency of the grid. The "Danish Concept," of directly connecting a wind turbine to the grid, is widely used for power ratings up to 2.3 MW. The scheme consists of a squirrel-cage induction generator (SCIG), connected via a transformer to the grid. The wind turbine systems using cage rotor induction generators almost operate at a fixed speed (variation of 1–2%). The power can be limited aerodynamically by stall control, active stall control, or by pitch control. The basic configurations of three different fixed speed concepts are shown in Fig. 3.1. The advantage of wind turbines with induction generators is the simple and cheap construction. In addition, no synchronization device is required. These systems are attractive due to cost and reliability, but they are not fast enough (within a few ms) to control the active power. There are some other drawbacks also: the wind turbine has to operate at constant speed, it requires a stiff power grid to enable stable operation, and it may require a more expensive mechanical construction in order to absorb high mechanical stress since wind gusts may cause torque pulsations in the drive train and the gearbox. Other disadvantages with the induction generators are high starting currents and their demand for reactive power. They need a reactive power compensator to reduce (almost eliminate) the reactive power demand from the turbine generators to the grid. It is usually done by continuously switching capacitor banks following the production variation (5–25 steps).

Connecting the induction generators to power system produces transients that are short duration, very high inrush currents causing both disturbances to the grid and high torque spikes in the drive train of wind turbines with a directly connected induction generator.

Unless special precautions are taken, the inrush currents can be up to 5–7 times the rated current of the generator; however, after a very short period (less than 100 ms), the current peak may be considerably higher, up to 18 times the normal rated current. A transient like this disturbs the grid and limits the acceptable number of value of all wind turbines. All three systems shown in Fig. 3.1 use a thyristor controller, the soft starter (not shown in Fig. 3.1), in order to reduce the inrush current [10]. The current limiter, or soft starter, based on thyristor technology, typically limits the highest rms value

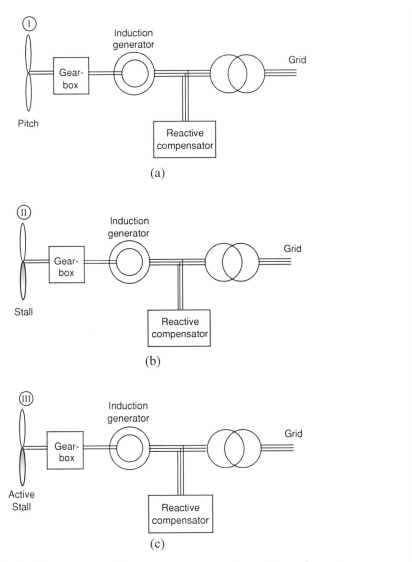

FIGURE 3.1: Wind turbine systems without power converter, but with aerodynamic power control. (a) Pitch controlled (System I), (b) stall controlled (System II), and (c) active stall controlled (System III).

of the inrush current to a level that is two times below that of the generator rated current. The soft starter has a limited thermal capacity and so it is short circuited by a contactor, which carries the full load current when the connection to the grid has been completed. In addition to reducing the impact on the grid, the soft starter also effectively dampens the torque peaks associated with the peak currents and hence reduces the loads on the gearbox.

An example is shown here to illustrate the startup of a soft-starter-fed induction generator [11]. The induction machine has 2 MW rated power, 690 V/1700 A rated phase voltage and rated line current, respectively (delta connection). The induction machine is connected via a soft starter to the supply voltage below synchronous speed (1450 rpm). The starting firing angle for the soft starter is 120°. The equivalent diagram of this system is shown in Fig. 3.2(a). The electromagnetic torque and the rotational speed of the high-speed shaft during the startup are presented in two cases: direct startup and using a soft starter. Fig. 3.2(b) shows the simulation results for the direct startup, while Fig. 3.2(c) shows the results when the machine is connected to the grid via a soft starter. When the induction machine is connected directly to the grid, high starting torque is observed. Large oscillations in the shaft speed can be seen in Fig. 3.2(b). By using a soft starter, the inrush currents and therefore the high starting torque are limited and the shaft speed is smoothed as shown in Fig. 3.2(c).

## 3.2    VARIABLE-SPEED WIND TURBINES

In variable-speed systems the generator is normally connected to the grid by a power electronic system. For synchronous generators and for induction generators without rotor windings, a full-rated power electronic system is connected between the stator of the generator and the grid, where the total power production must be fed through the power electronic system [12, 13]. For induction generators with rotor windings, the stator of the generator is connected to the grid directly. Only the rotor of the generator is connected through a power electronic system. This gives the advantage that only a part of the power production is fed through the power electronic converter. This means the nominal power of the converter system can be less than the nominal power of the wind turbine. In general the nominal power of the converter may be 30% of the power rating of the wind turbine, enabling a rotor speed variation in the range of 30% of the nominal speed. By controlling the active power of the converter, it is possible to vary the rotational speed of the generator and thus of the rotor of the wind turbines.

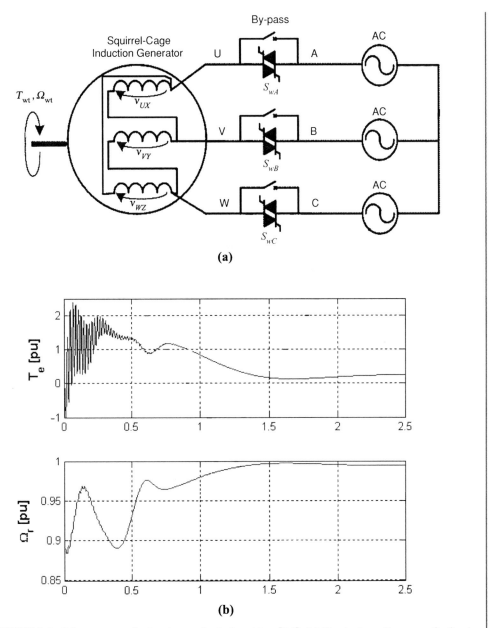

FIGURE 3.2:  The startup of a fixed-speed wind turbine [11]. (a) Equivalent diagram of a fixed-speed wind turbine to show the startup. (b) Electromagnetic torque and shaft speed during the direct startup of a 2 MW induction machine. (c) Electromagnetic torque and shaft speed during startup of a 2 MW soft-starter-fed induction machine.

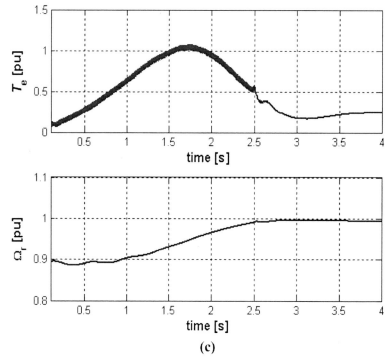

**FIGURE 3.2:** *(Continued)*

## 3.2.1 Variable-Speed Wind Turbines with Partially Rated Power Converters

The next category is wind turbines with partially rated power converters. By using these wind turbines the improved control performance can be obtained. Fig. 3.3 shows two such systems [14, 15]. The generator for wind turbine systems shown in Fig. 3.3 is an induction generator with a wounded rotor.

### 3.2.1.1 Dynamic Slip-Controlled Wounded Rotor Induction Generator

In Fig. 3.3(a) an extra resistance is added in the rotor, which can be controlled by power electronics. The variation of rotor resistance produces a group of torque-speed characteristics as shown in Fig. 3.4. This is known as the dynamic slip control and gives typically a speed range of 2–5%. The power converter for the rotor resistance control is for low voltage but high currents. At the same time an extra control freedom is obtained

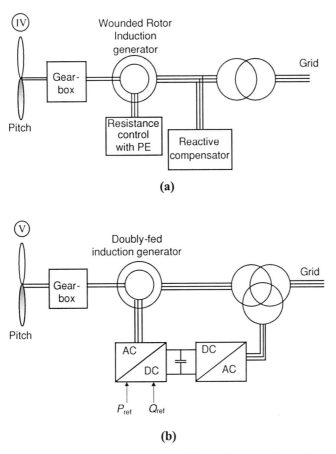

**(a)**

**(b)**

FIGURE 3.3: Wind turbine topologies with partially rated power electronics and limited speed range. Rotor-resistance converter (System IV) and doubly-fed induction generator (System V).

at higher wind speeds in order to keep the output power fixed. This system still needs a soft starter and reactive power compensation.

### 3.2.1.2 Doubly Fed Induction Generator

A doubly fed induction generator (DIFG) using a medium scale power converter is shown in Fig. 3.3(b). Slip rings are making the electrical connection to the rotor. If the generator is running super-synchronously, electrical power is delivered to the grid through both the rotor and the stator. If the generator is running sub-synchronously, electrical power is delivered into the rotor from the grid. A speed variation of $\pm 30\%$

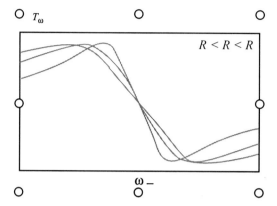

FIGURE 3.4: Torque and speed characteristics of rotor resistance controlled wound rotor induction generator.

around synchronous speed can be obtained by the use of a power converter of 30% of nominal power. Furthermore, it is possible to control both active ($P_{ref}$) and reactive power ($Q_{ref}$), which gives a better grid performance, and the power electronics enables the wind turbine to act as a more dynamic power source to the grid.

The DFIG system does not need either a soft starter or a reactive power compensator. The system is naturally a little bit more expensive compared to the classical systems shown before in Figs. 3.1 and 3.3(a). However, it is possible to save money on the safety margin of gear and reactive power compensation units, and it is also possible to capture more energy from the wind.

### 3.2.2   Full Scale Power Electronic Converter Integrated Systems

The wind turbines with a full-scale power converter between the generator and the grid give the added technical performance. Fig. 3.5 shows four possible systems with full-scale power converters.

The systems shown in Figs. 3.5(a) and 3.5(b) are characterized by having a gearbox. The wind turbine system with a cage rotor induction generator and full-rated power electronic converters is shown in Fig. 3.5(a). Usually, a back-to-back voltage source converter is used in order to achieve full control of the active and reactive power.

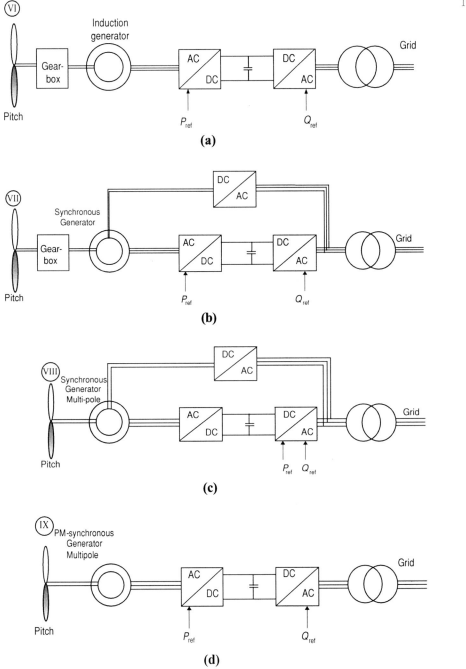

**FIGURE 3.5:** Wind turbine systems with full-scale power converters. (a) Induction generator with gear (System VI). (b) Synchronous generator with gear (System VII). (c) Multipole synchronous generator (System VIII). (d) Multipole permanent magnet synchronous generator (System IX).

The synchronous generator shown in Fig. 3.5(b) needs a small power converter for field excitation. Multipole systems with the synchronous generator without a gear are shown in Figs. 3.5(c) and 3.5(d). The last system is using permanent magnets, which are becoming cheaper and thereby attractive. All four systems have almost the same controllable characteristics since the generator is decoupled from the grid by a dc link. The power converter to the grid enables the system to control active and reactive power very fast. However, the disadvantage is a more complex system with more sensitive electronic parts.

## 3.3    SUMMARY OF WIND TURBINE-GENERATOR SYSTEMS

By introducing power electronics many of the wind turbine systems perform like a power plant. With regards to control performance they are faster but of course the real power produced depends on the available wind. In some systems, the reactive power can be delivered without having any wind.

Table 3.1 presents a technical comparison of the above discussed wind turbine systems, where comments on grid control, cost, maintenance, internal turbine performance are given. Each of these topologies has benefits and drawbacks. A fixed-speed wind turbine is relatively simple, so the price tends to be slightly lower. Since the rotor speed cannot be varied, these turbines must be more robust than the other designs due to the higher structural loads involved.

A variable-speed wind turbine generates more energy for a given wind speed, especially at low wind speed. Moreover, the active and reactive power can be easily controlled and there is less mechanical stress. The direct-driven machines are normally more expensive. However, costs can be saved on the gearbox. The major drawback of the direct-driven topologies is the large and relatively heavy generator. Moreover, the power converter has to be designed to handle the full-rated power.

Table 3.1 also lists some other important issues for wind turbines acting as a real power source in the grid. The rolling capacity on grid means that a wind turbine keeps active when a fault appears in the grid and has the possibility to reduce the power

TABLE 3.1: Comparison of Different Wind Turbine Systems

| | SYSTEM | | | | | | | | |
|---|---|---|---|---|---|---|---|---|---|
| | I | II | III | IV | V | VI | VII | VIII | IX |
| Variable speed | No | No | No | No | Yes | Yes | Yes | Yes | Yes |
| Control active power | Limited | No | Limited | Limited | Yes | Yes | Yes | Yes | Yes |
| Control reactive power | No | No | No | No | Yes | Yes | Yes | Yes | Yes |
| Short circuit (fault-active) | No | No | No | No | No/Yes | Yes | Yes | Yes | Yes |
| Short circuit power | Contribute | Contribute | Contribute | Contribute | Contribute | Limit | Limit | Limit | Limit |
| Control bandwidth | 1–10 s | 1–10 s | 1–10 s | 100 ms | 1 ms | 0.5–1 ms | 0.5–1 ms | 0.5–1 ms | 0.5–1 ms |
| Standby function | No | No | No | No | Yes + | Yes ++ | Yes ++ | Yes ++ | Yes ++ |

(Continued)

TABLE 3.1: (*Continued*)

| | SYSTEM | | | | | | | | |
|---|---|---|---|---|---|---|---|---|---|
| | I | II | III | IV | V | VI | VII | VIII | IX |
| Flicker (sensitive) | Yes | Yes | Yes | Yes | No | No | No | No | No |
| Soft starter needed | Yes | Yes | Yes | Yes | No | No | No | No | No |
| Rolling capacity on grid | Yes, partly | No | Yes, partly | Yes, partly | Yes | Yes | Yes | Yes | Yes |
| Reactive compensator (C) | Yes | Yes | Yes | Yes | No | No | No | No | No |
| Island operation | No | No | No | No | Yes/No | Yes/No | Yes/No | Yes/No | Yes |
| Investment | ++ | ++ | ++ | ++ | + | 0 | 0 | 0 | 0 |
| Maintenance | ++ | ++ | ++ | ++ | 0 | + | + | + | + |

++ Very positive (low cost); +, positive; 0, not competitive in cost.

TABLE 3.2: Wind Turbine Topologies Market in 2001

| TURBINE CONCEPT | MARKET SHARE (%) | |
| --- | --- | --- |
| | WORLD | GERMANY |
| Fixed speed (stall or active stall, gearbox), System I, II, III | 23 | 22 |
| Dynamic slip control (limited variable speed, pitch control, gearbox), System IV | 11 | 0 |
| Doubly fed generator (variable speed operation, pitch control, gearbox), System V | 50 | 49 |
| Direct-driven (variable speed operation, pitch control), System VIII | 16 | 29 |
| **Total** | **100** | **100** |

*Source:* BTM consult [16].

production even though more power is available in the wind and thereby act as a rolling capacity for the power system. The island operation may occur in some situations in case of a grid collapse and the ability of operating in such a situation could be useful for minimizing the possible area of blackout. The market share in 2001 (globally and in Germany), between the dominant system topologies, is shown Table 3.2. As can be seen the most sold technology in 2001 was the doubly-fed generator system, which occupies about 50% of the whole market. More than 75% of all sold wind turbines in 2001 were controlled by power electronics. That is even more than in 2003.

CHAPTER 4

# Control of Wind Turbines

Overall, the power can be controlled by means of the aerodynamic system and has to follow a set point given by a dispatch center or locally, with the goal to maximize the production based on the available wind power. The power control system should also be able to limit the power. Controlling a wind turbine involves both fast and slow control. In this chapter we discuss some typical control methods associated with different wind power generators.

## 4.1   ACTIVE STALL WIND TURBINE WITH CAGE ROTOR INDUCTION GENERATORS

In principle, an active stall wind turbine is a stall turbine with a variable pitch angle. The main difference between a stall turbine and an active stall turbine is a pitch system for variable pitch angles, which allows the stall effect to be controlled. An active stall wind turbine has to pitch in a negative direction to limit the power when the electrical power of the wind turbine exceeds nominal power. The active stall system basically maintains all the characteristics of a stall-regulated system. Large wind farms such as Nysted (170 MW installed capacity) have been built with active stall wind turbines.

The generator of an active stall turbine can be a simple squirrel cage induction generator directly connected to the grid. In order to compensate for the output power factor, a capacitor bank is used. A soft starter is used only during the startup sequence of the generator in order to limit the inrush currents and hence reduce the high starting torque.

The maximum power output of the active stall turbines can be maintained at a constant value. In addition, the aerodynamic efficiency $C_p$ can be optimized to a certain extent. The active stall control can improve the efficiency of the overall system. The flexible coupling of the blades to the hub also facilitates emergency stopping and start up. One drawback of the active stall controlled wind turbine compared to the passive stall one is the higher price, which is due to the pitching mechanism and its controller.

The implemented active stall wind turbine controller achieves good power yield with a minimum of pitch actions. Once the overall mean wind speed is at a constant level, pitch angle adjustments are rarely necessary, allowing the controller to optimize the pitch angle as often as possible.

Depending on the pitch system, the lost power (due to slow control) may be justified by reduced stress and wear in the pitch system and reduced fatigue loads on the wind turbine. This applies both to power optimization, where the controller strives for maximum power yield by using the moving average of the wind speed signal to find the appropriate pitch angle in a lookup table, and to power limitation where the power output is controlled in a closed control loop. With a slow control system, substantial over-power in the power limitation mode may cause a problem. This may be avoided by an over-power protection feature.

**An example**

In Fig. 4.1, the following conditions are illustrated: the mean wind speed is 11 m/s until the simulation time reaches 60 s, and between 60 and 160 s it is ramped up from 11 to 16 m/s. This corresponds to a slope in mean wind speed of 3 m/s min.

The 2 MW turbine starts off in power optimization mode when an increase in pitch angle takes place. As the wind speed increases, the turbine enters power limitation mode; a further increase in the wind speed causes the average power to exceed 2300 kW (300 kW beyond nominal power, which is the maximum allowed level of tolerable over-power). As soon as the over-power is detected, the pitch angle is adjusted. For steadily increasing wind speed as shown in the figure, it takes three adjusting operations to finally limit the power at the nominal level as shown in Fig. 4.1.

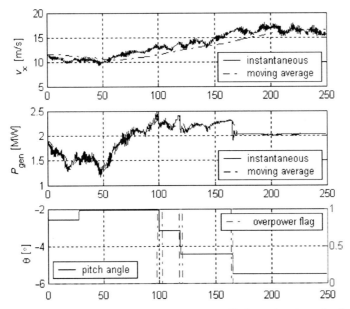

**FIGURE 4.1:** Simulation results for an active stall controlled wind turbine [11].

## 4.2   VARIABLE PITCH ANGLE CONTROL WITH DOUBLY FED GENERATORS

The variable speed DFIG wind turbine is a widely used concept today. The control system of a variable speed wind turbine with DFIG mainly functions to

- control the power drawn from the wind turbine in order to track the wind turbine optimum operation point,

- limit the power in the case of high wind speeds, and

- control the reactive power exchanged between the wind turbine generator and the grid.

Two hierarchical control levels are related to each other with different bandwidths, namely, DFIG control level and wind turbine control level. An example of an overall control scheme of a wind turbine with a doubly fed generator system is shown in Fig. 4.2.

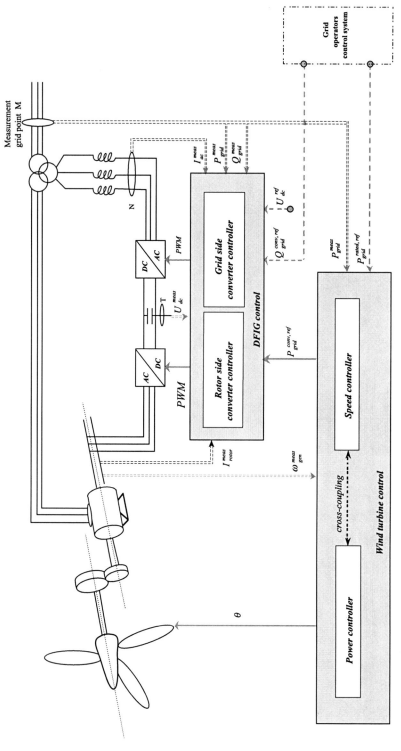

FIGURE 4.2: Control of wind turbine with DFIG system [17].

The DFIG control, with a fast dynamic response, contains the electrical control of the power converters and of the DFIG. The wind turbine control, with slow dynamic response, supervises both the pitch system of the wind turbine as well as the active power set point of the DFIG control level.

A vector control approach is adopted for the DFIG control, while two cross-coupled controllers are used to control the wind turbine. These controllers are speed and power limitation controllers. Their goals are to track the wind turbine optimum operation point, to limit the power in the case of high wind speeds, and to control the reactive power exchanged between the wind turbine generator and the grid.

Below maximum power production, the wind turbine will typically vary the speed proportionally with the wind speed and keep the pitch angle $\theta$ fixed. At very low wind, the speed of the turbine will be fixed at the maximum allowable slip in order not to have overvoltage. A pitch angle controller will limit the power when the turbine reaches the nominal power. The generated electrical power is controlled by the doubly fed generator through the rotor-side converter. The control of the grid-side converter simply keeps the dc-link voltage fixed. Internal current loops in both converters are used with typical PI-controllers. The power converters to the grid-side and the rotor-side are voltage source inverters.

The significant feature of the control method is that it allows the turbine to operate with optimum power efficiency over a wide range of wind speeds. Moreover, because of the design of this control method, small changes in generator speed do not lead to large power fluctuations and unnecessary transitions between power optimization and power limitation modes. A gain scheduling control of the pitch angle is also implemented in order to compensate for the nonlinear aerodynamic characteristics.

**An example**

The example below illustrates the performance of both the DFIG controller and the overall control of the variable speed/variable pitch wind turbine. The variable speed wind turbine has a rated power of 2 MW. The rated wind speed is at 11.5 m/s and the rated generator speed is 1686 rpm. Fig. 4.3 shows the simulation results when wind

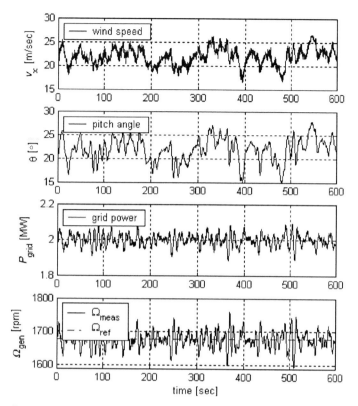

**FIGURE 4.3:** Simulation results for a variable speed/pitch wind turbine with DFIG with turbulent wind speed [11].

speeds, are at a mean value of 22 m/s, and with a turbulence intensity of 10%. Typical quantities, including wind speed, the pitch angle, the generator power to the grid, generator speed, and the reference of the generator speed, are shown in Fig. 4.3.

The wind speed corresponds to the power limitation strategy, where both the speed control loop and power control loop are active. The power control loop is fast, while the speed control loop is much slower, allowing dynamic variations of the generator speed in a predefined range. It can be seen that the power on the grid is limited to 2 MW, its variations being less than 2% of the rated power. The reference of the generator is maintained at the nominal speed, while the generator speed varies with the electrical power. The pitch angle reacts to the slow variations in the wind speed.

## 4.3    FULL RATED POWER ELECTRONIC INTERFACE WIND TURBINE SYSTEMS

Cage induction generators and synchronous generators can be integrated into the system by full rated power electronic converters. As shown in Fig. 4.4, a passive rectifier and a boost converter are used in order to boost the voltage at low speed [8, 9]. It is possible to control the active power from the generator. A grid inverter interfaces the dc-link to the grid. Here it is also possible to control the reactive power to the grid. The system is able to control reactive and active power quickly and then the turbine may take part in the power system control.

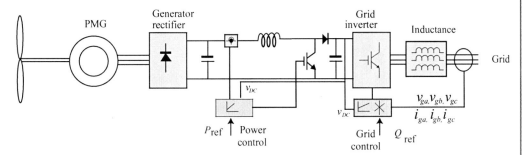

FIGURE 4.4: Basic control of active and reactive power in a wind turbine with a multipole synchronous generator system [8].

C H A P T E R   5

# Electrical Topologies of Wind Farms Based on Different Wind Turbines

In many countries, energy planning with a high penetration of wind energy is going on, which includes large wind farms. These wind farms may present a significant power contribution to the national grid, and therefore, play an important role in power quality and the control of power systems.

Consequently, high technical demands are expected to be met by these generation units in order to perform frequency and voltage control, the regulation of active and reactive power, and quick responses under power system transient and dynamic situations. For example, it may be required to reduce the power from the nominal power to 20% power within 2 s. The power electronic technology is again an important part in both the system configurations and the control of the wind farms in order to fulfill these demands. Also, the overall performance of a wind farm will largely depend on the types of the wind turbines installed as well as the topology of the electrical system. Some possible electrical configurations of wind farms are shown in Fig. 5.1. The requirements of integrating wind turbines and wind farms into power system are discussed in the next section.

A wind farm equipped with power electronic converters, as shown in Fig. 5.1(a), can perform both real and reactive power control and also operate the wind turbines in variable speed to maximize the energy captured as well as reduce the mechanical stress

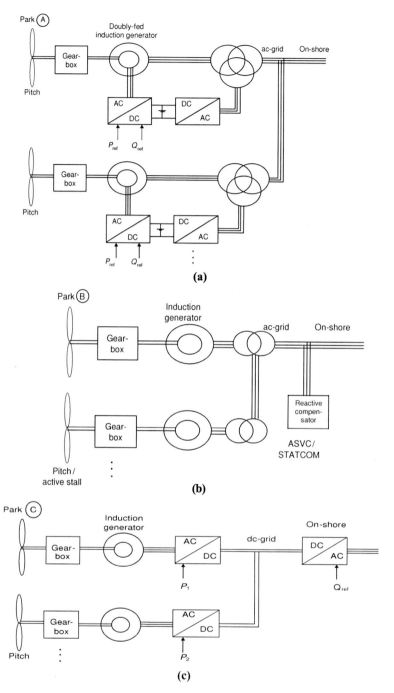

**FIGURE 5.1:** Wind farm solutions. (a) DFIG system with ac-grid (System A). (b) Induction generator with ac-grid (System B). (c) Speed controlled induction generator with common dc-bus and control of active and reactive power (System C). (d) Speed controlled induction generator with common ac-grid and dc transmission (System D).

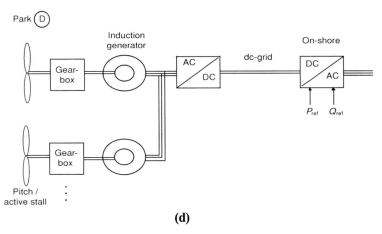

**(d)**

FIGURE 5.1: *(Continued)*

and noise. Such a system is in operation in Denmark as a 160 MW off-shore wind power station.

Fig. 5.1(b) shows a wind farm with induction generators. A STATCOM can be used to provide the reactive power control to meet the system reactive power control requirements, and it can help to control the voltage, as well as, provide the reactive power demand of the induction generators in the wind farm.

For long distance transmission of power from off-shore wind farms, HVDC may be an interesting option. In a HVDC transmission, the low or medium ac voltage at the wind farm is converted into a high dc voltage on the transmission side and the dc power is transferred to the onshore system where the dc voltage is converted back into ac voltage as shown in Fig. 5.1(d). For certain power level, a HVDC transmission system, based on voltage source converter technology, may be used in such a system instead of the conventional thyristor-based HVDC technology. The topology may even be able to vary the speed on the wind turbines in the complete wind farm. Another possible dc transmission system configuration is shown in Fig. 5.1(c), where each wind turbine has its own power electronic converter and so it is possible to operate each wind turbine at an individual optimal speed. A comparison of the topologies of these four wind farms is given in Table 5.1. As it can be seen, the wind farms have interesting

**TABLE 5.1:** Comparison of Four Wind Farm Topologies

| FARM CONFIGURATIONS | PARK A | PARK B | PARK C | PARK D |
|---|---|---|---|---|
| Individual speed control | Yes | No | Yes | No |
| Control active power electronically | Yes | No | Yes | Yes |
| Control reactive power | Yes | Centralized | Yes | Yes |
| Short circuit (active) | Partly | Partly | Yes | Yes |
| Short circuit power | Contribute | Contribute | No | No |
| Control bandwidth | 10–100 ms | 200 ms to 2 s | 10–100 ms | 10 ms to 10 s |
| Standby-function | Yes | No | Yes | Yes |
| Softstarter needed | No | Yes | No | No |
| Rolling capacity on grid | Yes | Partly | Yes | Yes |
| Redundancy | Yes | Yes | No | No |
| Investment | + | ++ | + | + |
| Maintenance | + | ++ | + | + |

++ cheaper, + more expensive

features so as to act as a power source to the grid. Some have better abilities than others. The overall considerations will include production, investment, maintenance, and reliability.

There are other possibilities, such as field excited synchronous machines or permanent magnet synchronous generators, that may be used in the systems shown in Fig. 5.1(c) or 5.1(d). In the case of a multipole generator, the gearbox may be removed.

CHAPTER 6

# Integration of Wind Turbines into Power Systems

Large-scale integration of wind turbines may have significant impacts on power system operation. Traditionally, wind turbines are not required to participate in frequency and voltage control. However, in recent years, attention has been increased on wind farm performance in power systems [18–22]. Consequently, some grid codes have been defined to specify the requirements that wind turbines must meet in order to be connected to the grid. Examples of such requirements include the capability of contributing to frequency and voltage control by continuously adjusting active power and reactive power supplied to the transmission system, and the power regulation rate that a wind farm must provide. Some of the requirements can be dealt with by implementing control schemes in certain types of wind turbines, such as reactive power control with wind turbines having power electronic converters. Many research activities have been conducted in this area [23–36]. In this chapter we describe main requirements, turbine performance assessments, and some possible methods for meeting the requirements.

## 6.1   REQUIREMENTS OF WIND TURBINE GRID INTEGRATION

### 6.1.1   Frequency and Active Power Control

The electrical supply and distribution systems used worldwide today are based on ac systems (50 or 60 Hz). The frequency of a power system is proportional to the rotating speed of the synchronous generators operating in the system. The generators in the same ac system are synchronized, running at the same speed. Increasing the electrical load in the system tends to slow down the generators and reduce the frequency. The task of frequency control of the system is to increase or reduce the generated power so as to keep the generators operating in the specified frequency range. However, renewable resources can only produce when the source is available. For wind power, this is when and where the wind blows. This characteristic is important when the amount of wind power covers a large fraction of the total demand for electricity energy in the system. In order to be able to increase the power output for frequency control, a wind turbine may have to operate at a lower power level than the available power, which means low utilization of the wind energy resources. One way to improve the situation may be the use of "energy storage" technologies, such as batteries, pump storage, and fuel cells, though the speed of response will vary depending on the energy storage technology. So far large-scale, cost-effective energy storage technologies are yet to be developed.

### 6.1.2   Short Circuit Power Level and Voltage Variations

The short circuit power level at a given point in an electrical network is a measure of its strength, and although it is not a direct parameter of voltage quality, it has a significant influence. The ability of the grid to absorb disturbances is directly related to the short circuit power level.

Considering a point in the network, the voltage far away from the point may not be influenced by the conditions at this point. $Z_k$ is the equivalent impedance between the concerned point and the remote location. $U_k$, is a nominal voltage of the point, and the short circuit power level $S_k$ in MVA can be found as $U_k^2 / Z_k$. Strong and/or weak grids are terms often used in connection with wind power installations. If the impedance

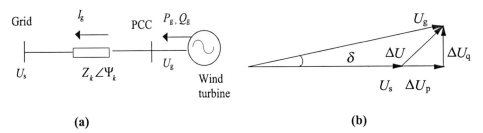

**FIGURE 6.1:** A simple system with an equivalent wind power generator connected to a network. (a) System circuit and (b) phasor diagram.

$Z_k$ is small, the voltage variations at PCC will be small (the grid is strong), but if $Z_k$ is large, the voltage variations will be large (the grid is weak).

Fig 6.1 illustrates an equivalent wind power generation unit, connected to a network with short circuit impedance $Z_k$. The network voltage at the assumed remote busbar and the voltage at the point of common coupling (PCC) are $U_s$ and $U_g$, respectively. The output power and reactive power of the generation unit are $P_g$ and $Q_g$, which correspond to a current $I_g$:

$$I_g = \left(\frac{S_g}{U_g}\right)^* = \frac{P_g - jQ_g}{U_s}. \tag{1}$$

The voltage difference $\Delta U$ between the system and the connection point is given by

$$U_g - U_s = \Delta U = Z_k I_g = (R_k + jX_k)\left(\frac{P_g - jQ_g}{U_g}\right)$$

$$= \frac{R_k P_g + X_k Q_g}{U_g} + j\frac{P_g X_k - Q_g R_k}{U_g} = \Delta U_p + j\Delta U_q. \tag{2}$$

$\Delta U$ is related to the short circuit impedance, the real and reactive power output of the wind power generation unit. It is clear that the variations of the generated power will result in variations of the voltage at PCC.

Equation (2) indicates the relationship between the voltage and the power transferred into the system. $\Delta U$ can be calculated with load flow methods as well as with other simulation techniques. The voltage at PCC should be maintained within utility regulatory limits. Operation of wind turbines may affect the voltage in the connected network. If necessary, appropriate precautions should be taken to ensure that the wind turbine installation does not bring the magnitude of the voltage outside the required limits.

## 6.1.3   Reactive Power Control

Conventional reactive power concept is associated with the oscillation of energy stored in capacitive and inductive components in a power system. Reactive power is produced in capacitive components and consumed in inductive components. A synchronous generator can either produce or consume reactive power by controlling the magnetizing level of the generator, i.e. a high magnetizing level results in high voltage and production of reactive power.

The current associated with the reactive power flow causes system voltage drop as aforementioned and also power losses. Furthermore, large reactive currents flowing in a power system may cause voltage instability in the network due to the associated voltage drops in the transmission lines. Therefore, reactive power control is important. The induction generator based wind turbines are the consumer of reactive power. To minimize the power losses and to increase voltage stability, these wind turbines are compensated to a level depending on the requirements of the local utility or distribution company. For wind turbines with PWM converter systems, the reactive power can be controlled by the converter. For example, these wind turbines can have a power factor of 1.00 and also have the possibility to control voltage by controlling the reactive power (generation or consumption of reactive power).

## 6.1.4   Flicker

Voltage variations caused by fluctuating wind power generation may cause voltage quality problems. Fluctuations in the system voltage (in terms of rms value) may cause perceptible light flicker depending on the magnitude and frequency of the fluctuation. This type of disturbance is called voltage flicker, commonly known as flicker.

The allowable flicker limits are generally established by individual utilities. Rapid variations in the power output from a wind turbine, such as generator switching and capacitor switching, can also result in variations in the rms value of the voltage. At certain rate and magnitude, the variations cause flickering of the electric light. In order to prevent flicker emission from impairing the voltage quality, the operation of the generation units should not cause excessive voltage flicker.

Flicker evaluation based on IEC 1000-3-7 gives guidelines for emission limits of fluctuating loads in medium voltage and high voltage networks. The basis for the evaluation is a measured curve giving the threshold of visibility for rectangular voltage changes applied to an incandescent lamp. The level of flicker is quantified by the short-term flicker severity $P_{st}$, which is normally measured over a 10-min period. Disturbances just visible are said to have a flicker severity factor of $P_{st} = 1$. Furthermore, a long-term flicker severity factor $P_{lt}$ is defined where $P_{lt}$ is measured over 2-h periods.

Determination of flicker emission can be done on the basis of measurement. IEC 61000-4-15 specifies a flickermeter, which can be used to measure flicker directly. The flicker emissions may be estimated with the coefficient and factors, $c_f (\Psi_k, v_a)$ and $k_f(\Psi_k)$ obtained from the measurements, which are usually provided by wind turbine manufacturers [23].

## 6.1.5   Harmonics

Harmonics are a phenomenon associated with the distortion of the voltage and current waveforms. Any periodical function may be expressed as a sum of sinusoidal waveforms with different frequencies including the fundamental frequency and a series of integer multiples of the fundamental component. Depending on the harmonic order these may cause damage of various kinds to different type of electrical equipment. All harmonics cause increased currents and possible destructive overheating in capacitors as the impedance of a capacitor goes down in proportion to the increase in frequency. The higher harmonics may further give rise to increased noise in analogue telephone circuits. The harmonic distortion is expressed as total harmonic distortion (THD). THD and individual harmonics should meet the system requirements.

The pulse width modulation (PWM) switching converters are used in most variable speed wind turbine technologies today. The switching frequency is typically around a few kilohertzs. The high-frequency harmonics are small in magnitude and are easier to be removed by filters.

## 6.1.6   Stability

The problem of network stability is often associated with different types of faults in the network, such as tripping of transmission lines (e.g. overload), loss of production capacity, and short circuits. Tripping of transmissions lines due to overload or component failure disrupts the balance of power (active and reactive) flow. Although the capacity of the operating generators may be adequate, large voltage drops may occur suddenly. The reactive power flowing through new paths in a highly loaded transmission grid may force the voltage of the network in the area down beyond the border of stability. Often a period of low voltage is followed by complete loss of power. Loss of production capacity obviously results in a large, momentary, power imbalance. Unless the remaining operating power plants have enough "spinning reserve," that is, generators are not loaded to their maximum capacity, to replace the loss within very short time, a large frequency and voltage drop will occur, followed by complete loss of power. One way of dealing with this situation is to disconnect the supply to some areas or some large consumers, so as to restore the power balance and to limit the number of consumers affected by the fault.

Short circuits have a variety of forms, from the one-phase earth fault caused by trees, to the three-phase short circuit with low impedance in the short circuit path. Many of these faults are cleared by the relay protection of the transmission system, either by disconnection and fast reclosure, or by disconnection of the equipment in question after a few hundred milliseconds. In all the situations, the result is a short period with low or no voltage followed by a period when the voltage restores. A large wind farm in the vicinity will see this event and may be disconnect from the grid if no appropriate control has been implemented. This leads to the situation "loss of production capacity." The disconnection of the wind farm will further aggravate the situation and therefore, in some grid codes, wind turbines and wind farms are required to have the ability of ride through. Studies show that different wind turbines may have different control methods during the transients [14, 27–36].

## 6.2    VOLTAGE QUALITY ASSESSMENT

The assessment of the impacts from integrating wind turbines may be performed according to the methods given in the IEC 61400-21 /2/ to determine the acceptability of such integration. Methods include:

- steady-state voltage

- flicker

- harmonics

### 6.2.1    Steady-State Voltage

The grid and wind turbine voltage should be maintained within the utility limits. Operation of a wind turbine may affect the steady-state voltage in the network. It is recommended that load-flow analyses be conducted to assess this effect to ensure that the wind turbine installation does not bring the magnitude of the voltage beyond the required limits of the network. In general, some extreme case of the loads and the wind turbine production may be checked for compatibility, such as

- low loads and low wind power,

- low loads and high wind power,

- high loads and low wind power, and

- high loads and high wind power.

Depending on the scope of the load-flow analysis, a wind turbine installation may be assumed as a PQ node, which may use 10-min average data ($P_{mc}$ and $Q_{mc}$) or 60-s average data ($P_{60}$ and $Q_{60}$) or 0.2-s average data ($P_{0.2}$ and $Q_{0.2}$).

A wind farm with multiple wind turbines may be represented with its output power at the PCC. Ten-minute average data ($P_{mc}$ and $Q_{mc}$) and 60-s average data ($P_{60}$ and $Q_{60}$) can be calculated by simple summation of the output from each wind turbine, whereas 0.2-s average data ($P_{0.2}$ and $Q_{0.2}$) may be calculated according to (3) and (4):

$$P_{0.2\Sigma} = \sum_{i=1}^{N_{wt}} P_{n,i} + \sqrt{\sum_{i=1}^{N_{wt}} (P_{0.2,i} - P_{n,i})^2} \qquad (3)$$

$$Q_{0.2\Sigma} = \sum_{i=1}^{N_{wt}} Q_{n,i} + \sqrt{\sum_{i=1}^{N_{wt}} (Q_{0.2,i} - Q_{n,i})^2} \qquad (4)$$

where $P_{n,i}$ and $Q_{n,i}$ are the rated real and reactive power of the individual wind turbine and $N_{wt}$ is the number of wind turbines in the group.

## 6.2.2   Voltage Fluctuations

There are two types of flicker emissions: the flicker emission during continuous operation and the flicker emission due to generator and capacitor switchings. Often, one or the other will be predominant.

The flicker emissions from a wind turbine installation should be limited to comply with the flicker emission limits. However, different utilities may have different flicker emission limits. The assessments of the flicker emissions are described below.

### 6.2.2.1  Continuous Operation

The flicker emission from a single wind turbine during continuous operation may be estimated by

$$P_{st} = c_f(\psi_k, v_a) \frac{S_n}{S_k} \qquad (5)$$

where $c_f(\Psi_k, v_a)$ is the flicker coefficient of the wind turbine for the given network impedance phase angle $\Psi_k$ at the PCC and for the given annual average wind speed $v_a$ at hub-height of the wind turbine.

A table of data produced from the measurements at a number of specified impedance angles and wind speeds can be provided by wind turbine manufacturers. From the table, the flicker coefficient of the wind turbine for the actual $\Psi_k$ and $v_a$ at the site may be found by applying linear interpolation. The flicker emission from a group of wind turbines connected to the PCC is estimated by (6):

$$P_{st \, \Sigma} = \frac{1}{S_k} \sqrt{\sum_{i=1}^{N_{wt}} (c_{f,i}(\psi_k, v_a) S_{n,i})^2} \tag{6}$$

where $c_{f,i}(\Psi_k, v_a)$ is the flicker coefficient of the individual wind turbine, $S_{n,i}$ is the rated apparent power of the individual wind turbine, and $N_{wt}$ is the number of wind turbines connected to the PCC.

If the limits of the flicker emission are known, the maximum allowable number of wind turbines for connection can be determined.

### 6.2.2.2  Switching Operations

The flicker emission due to switching operations of a single wind turbine can be calculated as

$$P_{st} = 18 \times N_{10}^{0.31} \times k_f(\psi_k) \frac{S_n}{S_k} \tag{7}$$

where $k_f(\Psi_k)$ is the flicker step factor of the wind turbine for the given $\Psi_k$ at the PCC. The flicker step factor of the wind turbine for the actual $\Psi_k$ at the site may be found by applying linear interpolation to the table of data produced from the measurements by wind turbine manufacturers.

The flicker emission from a group of wind turbines connected to the PCC can be estimated from

$$P_{st \, \Sigma} = \frac{18}{S_k} \left( \sum_{i=1}^{N_{wt}} N_{10,i} \, (k_{f,i}(\psi_k) S_{n,i})^{3.2} \right)^{0.31} \tag{8}$$

where $N_{10,i}$ and $N_{120,i}$ are the number of switching operations of the individual wind turbine within 10-min and 2-h periods, respectively, $k_{f,i}(\Psi_k)$ is the flicker step factor of the individual wind turbine, and $S_{n,i}$ is the rated apparent power of the individual wind turbine.

Again, if the limits of the flicker emission are given, the maximum allowable number of switching operations in a specified period, or the maximum permissible flicker emission factor, or the required short circuit capacity at the PCC may be determined.

## 6.2.3 Harmonics

A wind turbine with an induction generator directly connected to the grid is not expected to cause any significant harmonic distortions during normal operation. Only wind turbines with power electronics need to be checked concerning harmonics. The harmonic current emission of such wind turbine system is normally given in the power quality data sheet, while the limits for harmonics are often specified for harmonic voltages. Thus harmonic voltages should be calculated from the harmonic currents of the wind turbine, which requires the information of the grid impedances at different frequencies.

**An example of voltage flicker compensation**

Voltage variation and flicker emission of grid connected wind turbines are related to many factors, including

- mean wind speed $v$
- turbulence intensity In, and
- short circuit capacity ratio SCR $= S_k/S_n$, where $S_k$ is the short circuit capacity of the grid where the wind turbines are connected and $S_n$ is the rated power of the wind turbine.

The voltage difference in (2) may be approximated as

$$\Delta U \approx \Delta U_p = \frac{P_g R_k + Q_g X_k}{U_g} \tag{9}$$

With the grid impedance angle $\psi_k$ and the wind turbine power factor angle $\psi$ being defined as

$$\tan \psi_k = X_k/R_k$$
$$\tan \psi = Q_g/P_g \tag{10}$$

Equation (9) can be written as

$$\Delta U_p = \frac{P_g R_k(1 + \tan \psi_k \cdot \tan \psi)}{U_g} = \frac{P_g R_k \cos(\psi - \psi_k)}{U_g \cos \psi_k \cdot \cos \psi} \tag{11}$$

It can be seen from (11) that when the difference between the grid impedance angle $\psi_k$ and the wind turbine power factor angle $\psi$ approaches 90°, the voltage fluctuation is minimized [23]. Equation (11) also indicates that the reactive power may be regulated with the real power generation to minimize voltage variation and flicker.

The variable speed wind turbine with DFIG is capable of controlling the output of active and reactive power, respectively. Normally the output reactive power of the wind turbine is controlled as zero to keep the unity power factor. It is possible to control the output reactive power appropriately with the variation of the output real power so that the voltage changes from the real power flow may be cancelled by the reactive power flow. As mentioned before, when the difference between the grid impedance angle $\psi_k$ and the line power factor angle $\psi$ approaches 90°, the flicker emission is minimized. Therefore, the reactive power can be controlled by the grid side converter in proportion to the wind turbine output active power such that the power factor angle $\psi$ is close to the value of $\psi_k + 90°$.

According to IEC standard IEC 61000-4-15, a flickermeter model is built to calculate the short-term flicker severity $P_{st}$ as shown in Fig. 6.2.

Fig. 6.3 shows the wind speed and output power produced by the simulation model and the corresponding power spectra are shown in Fig. 6.4, where the 3p effect can be clearly seen. A simulation study has been carried out to investigate the flicker

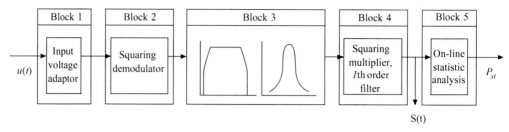

FIGURE 6.2: Flickermeter model according to IEC 61000-4-15.

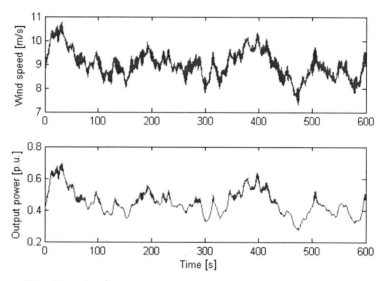

FIGURE 6.3: Wind speed and output power of a wind turbine [26].

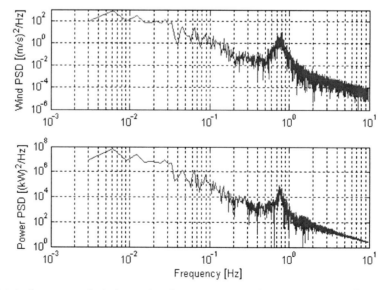

FIGURE 6.4: Spectrum of wind speed and output power of the wind turbine [26].

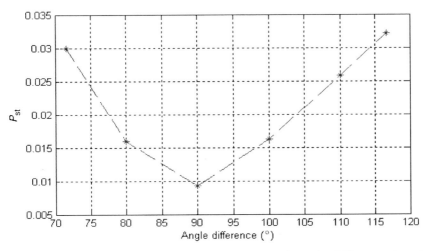

FIGURE 6.5: Short-term flicker severity $P_{st}$ variation with angle difference, $\psi - \psi_k$, $(v = 9 \text{ m/s}, \ln = 0.1, \text{SCR} = 20 \ \psi_k = 63.4°)$ [26].

minimization by using the reactive power generated from the power electronic converter at grid side of a DFIG system. The results, as shown in Fig. 6.5, indicate that the flicker level is significantly reduced if the angle difference $(\psi - \psi_k)$ is regulated to be 90° by controlling the reactive power flow.

# Conclusion

In this monograph we discussed the electrical aspects of wind turbine systems. Various wind turbine systems with different generators and power electronic converters have been described. Different types of wind turbine systems will have quite different performances and controllability, which was discussed with some results from study examples. The wind farms with different turbines may need different configurations for best use of the technical merits; therefore, electrical topologies of wind farms with different wind turbines have been briefed. Finally, main grid connection requirements, which are closely related to the electrical characteristics of the turbine systems, have been discussed. The possibility of using reactive power generated by the grid side converter of a doubly fed induction generator wind turbine systems to compensate voltage fluctuation has been illustrated.

# References

1. F. Blaabjerg, Z. Chen, and S. B. Kjaer, "Power electronics as efficient interface in dispersed power generation systems," *IEEE Trans. Power Electron.,* Vol. 19, No. 5, pp. 1184–1194, Sep. 2004.

2. Z. Chen and F. Blaabjerg, "Wind Turbines—A Cost Effective Power Source", Przeglad Elektrotechniczny R. 80 NR 5/2004 pp 464-469 (Journal, ISSN 0033-2097).

3. L. H. Hansen, P. H. Madsen, F. Blaabjerg, H. C. Christensen, U. Lindhard, and K. Eskildsen, "Generators and power electronics technology for wind turbines," in *Proc. IECON'01,* Vol. 3, pp. 2000–2005, 2001.

4. A. K. Wallace and J. A. Oliver, "Variable-speed generation controlled by passive elements," in *Proc. of ICEM'98,* 1998.

5. B. J. Baliga, "Power IC's in the saddle," *IEEE Spectrum,* pp. 34–49, July 1995.

6. Z. Chen and E. Spooner, "Current source thyristor inverter and its active compensation system," *IEE Proc. Generation, Transmission Distributions,* Vol. 150, No. 4, pp. 447–454, July 2003.

7. Z. Chen, "Compensation schemes for a SCR converter in variable speed wind power systems," *IEEE Trans. Power Delivery,* Vol. 19, No. 2, pp. 813–821, April 2004.

8. M. P. Kazmierkowski, R. Krishnan, and F. Blaabjerg, *Control in Power Electronics-Selected Problems.* Academic Press, New York, 2002.

9. Z. Chen and E. Spooner, "Voltage source inverters for high-power, variable-voltage DC power sources," *IEE Proc. Generation, Transmission and Distributions,* Vol. 148, No. 5, pp. 439–447, Sept 2001.

10. F. Iov, Z. Chen, F. Blaabjerg, A. Hansen, and P. Sorensen, "A new simulation platform to model, optimize and design wind turbine," in *Proc. of IECON'02,* Vol. 1, pp. 561–566.

11.  F. Iov, A. D. Hansen, C. Jauch, P. Sørensen, and F. Blaabjerg, "Advanced tools for modeling, design and optimization of wind turbine systems," *J. Power Electron.*, vol. 5, No. 2, 2005, pp 83–98.

12.  Z. Chen and E. Spooner, "Grid interface options for variable-speed, permanent-magnet generators," *IEE Proc. Electr. Power Applications*, Vol. 145, No. 4, pp. 273–283, Jul 1998.

13.  Z. Chen, S. Gómez Arnalte, and M. McCormick, "A fuzzy logic controlled power electronic system for variable speed wind energy conversion systems," in *8th IEE Int. Conf. PEVD'2000*, London, September 2000, pp. 114–119, IEE Conf. Publ. No. 475.

14.  T. Sun, Z. Chen, and F. Blaabjerg, "Voltage recovery of grid-connected wind turbines after a short-circuit fault," in *Proc. of the 29th Annual Conf. IEEE Industrial Electron. Soc., IECON 2003*, Roanoke, VA, 2003, pp. 2723–2728.

15.  R. Pena, J. C. Clare, and G. M. Asher, "Doubly fed induction generator using back-to-back PWM converters and its application to variable speed wind-energy generation," *IEE Proc. Electron. Power Appl.* pp. 231–241, 1996.

16.  BTM Consults Aps. "International wind energy department word market update 2002," Forecast 2003–2007, 2003.

17.  A. D. Hansen, C. Jauch, P. Soerensen, F. Iov, and F. Blaabjerg, "Dynamic wind turbine models in power system simulation tool DigSilent," Report Risoe-R-1400 (EN), Dec. 2003, ISBN 87-550-3198-6.

18.  IEC 61400-21, "Power quality requirements for wind whines," 2001.

19.  DEFU Committee reports 111-E, 2nd edition, "Connection of wind turbines to low and medium voltage networks 1998".

20.  IEC 61400-12, "Wind turbine generator systems. Power performance measurement techniques."

21.  IEC 61000-4-15, "Electromagnetic Compatibility (EMC)—Part 4: Testing and measurement techniques—Section 15: Flickermeter—Functional and design specifications," Bureau Central Commission Electrotech. Int., Geneva, Switzerland, Nov. 1997.

22. English version of Technical Regulations TF 3.2.6, "Wind turbines connected to grids with voltage below 100 kV—Technical regulations for the properties and the control of wind turbines," Eltra and Ekraft Systems, 2004.

23. Z. Chen, F. Blaabjerg, and T. Sun, "Voltage quality of grid connected wind turbines," *Proc. of the Workshop of Techniques and Equipments for Quality Ad Reliability of Electrical Power*, Bucharest, Romania, Printech, April 2004, pp. 11–16.

24. F. Blaabjerg and Z. Chen, "Wind power—A power source enabled by power electronics," *Proc. of 2004 CPES Power Electronics Seminar*, April 2004, pp. I3–I14.

25. Z. Chen and E. Spooner, "Grid power quality with variable-speed wind turbines," *IEEE Trans. Energy Conversion*, Vol. 16, No. 2, pp. 148–154, June 2001.

26. T. Sun, Z. Chen, and F. Blaabjerg, "Flicker study on variable speed wind turbines with doubly fed induction generators," *IEEE Trans. Energy Conversion*, Volume 20, Issue 4, Dec. 2005 pp. 896–905.

27. T. Sun, Z. Chen, and F. Blaabjerg, "Transient stability of DFIG wind turbines at an external short-circuit fault," *Wind Energy*, 2005, 8: 345–360.

28. Z. Saad-Saoud and N. Jenkins, "The application of advanced static VAr compensators to wind farms," *IEEE Colloquium on Power Electron. Renewable Energy*, (Digest No: 1997/170), 16 June 1997, pp. 6/1–6/5.

29. T. Petru and T. Thiringer, "Modelling of wind turbines for power system studies," *IEEE Trans. Power Systems*, Vol. 17, No. 4, pp. 1132–1139, Nov 2002.

30. J. G. Slotweeg, H. Polinder, and W. L. Kling, "Initialization of wind turbine models in power system dynamics simulations," in *IEEE Porto Power Tech*, Portugal, Sept. 10–13, 2001, pp. 6.

31. R. Flølo, M. Gustafsson, R. Fredheim, and T. Gjengedal, "Dynamic simulation of power systems with wind turbines, a case study from Norway," in *NWPC'00*, Trondheim, March 13–14, 2000, pp. 6.

32. A. Knudsen, "Modelling of windmill induction generators in dynamic simulation programs," in *Proc. IEEE Power Tech'99*, 1999, pp. 6.

33. Z. Chen and Y. Hu, "Dynamics performance improvement of a power electronic interfaced wind power conversion system," in *Proc. 4th Int. Power Electron. Motion Control Conf., IPEMC 2004*, Xi'an, August 2004.

34. Z. Chen and Y. Hu, "Power system dynamics influenced by a power electronic interface for variable speed wind energy conversion systems," in *Proc. 39th Int. Univ. Power Engineering Conf., UPEC 2004*, pp. 659–663.

35. J. Wiik, J. O. Gjerde, T. Gjengedal, and M. Gustafsson, "Steady state power system issues when planning large wind farms," in *IEEE PES 2002 Winter Meeting*, New York, 2002.

36. Z. Chen, F. Blaabjerg, and Y. Hu, "Voltage recovery of dynamic slip control wind turbines with a STATCOM," in *Proc. 2005 Int. Power Electronics Conf. IPEC 2005*, April 2005, pp. 1093–1100.

# The Authors

**Frede Blaabjerg** was born in Erslev, Denmark, on May 6, 1963. He received the M.Sc.EE. from Aalborg University, Denmark in 1987, and the Ph.D. degree from the Institute of Energy Technology, Aalborg University, in 1995.

He was employed at ABB-Scandia, Randers, from 1987 to 1988. During 1988–1992 he was a Ph.D. student at Aalborg University. In 1992, he became an Assistant Professor at Aalborg University, in 1996 an Associate Professor, and in 1998 Professor in power electronics and drives. In 2000, he was a Visiting Professor at the University of Padova, Italy, as well as part-time programme research leader at Research Center Risoe. In 2002, he was a Visiting Professor at Curtin University of Technology, Perth, Australia. His research areas include power electronics, static power converters, ac drives, switched reluctance drives, modelling, characterization of power semiconductor devices and simulation, power quality, wind turbines, and green power inverter. He is involved in more than fifteen research projects with the industry. Among these has been the Danfoss Professor Programme in Power Electronics and Drives. He is the author or coauthor of more than 500 publications in his research fields including the book *Control in Power Electronics* (edited by M.P. Kazmierkowski, R. Krishnan, F. Blaabjerg) published by Academic Press in 2002.

Dr. Blaabjerg is a member of the European Power Electronics and Drives Association and the IEEE Industry Applications Society Industrial Drives Committee. He is also a member of the Industry Power Converter Committee and the Power Electronics Devices and Components Committee in the IEEE Industry Application Society. He is the Associated Editor of the *IEEE Transactions on Industry Applications, IEEE Transactions on Power Electronics, Journal of Power Electronics*, and of the Danish journal *Elteknik*. He is also Editor-in-Chief of *IEEE Transactions on Power Electronics*.

He has served as member of the Danish Technical Research Council in Denmark from 1997 to 2003, and from 2001 to 2003 he was the chairman. He has also been the chairman of the Danish Small Satellite programme and the Center Contract Committee, which supports collaboration between universities and industry. He became a member of the Danish Academy of Technical Science in 2001 and in 2003 he became a member of the academic council. From 2002 to 2003 he was a member of the Board of the Danish Research Councils. From 2004 to 2005 he was the chairman of the programme committee Energy and Environment.

He has received the 1995 Angelos Award for his contribution in modulation technique and control of electric drives, and an Annual Teacher prize at Aalborg University, also 1995. In 1998 he received the Outstanding Young Power Electronics Engineer Award from the IEEE Power Electronics Society. He has received six IEEE Prize paper awards during the last six years and another prize paper award at PELINCEC Poland 2005. In 2002 he received the C.Y. O'Connor fellowship from Perth, Australia, in 2003 the Statoil-prize for his contributions in Power Electronics, and in 2004 the Grundfos Prize in acknowledgment of his international scientific research in power electronics. He became an IEEE Fellow in 2003.

**Zhe Chen** received his B.Eng. and M.Sc. from Northeast China Institute of Electric Power Engineering, China, and his Ph.D. from The University of Durham, UK. He was a Lecturer, then a Senior Lecturer with De Montfort University, UK. Since 2002, he has been a Research Professor with the Institute of Energy Technology (IET), Aalborg University, Aalborg, Denmark. Dr. Chen is the coordinator of the Wind Turbine Research Program at IET. His main research areas are renewable energy, power electronics, and power systems. He is the author, or co-author, of more than 100 publications in his research fields. He is a senior member of IEEE, a member of IEE (London), and a Chartered Engineer (UK).